PHONES
from Then to Now

BY ANGIE SMIBERT

SEQUENCE

AMICUS | AMICUS INK

Sequence is published by Amicus and Amicus Ink
P.O. Box 227, Mankato, MN 56002
www.amicuspublishing.us

Library of Congress Cataloging-in-Publication Data
Names: Smibert, Angie, author.
Title: Phones from then to now / by Angie Smibert.
Description: Mankato : Amicus, 2022. | Series: Developments in technology | Includes bibliographical references and index. | Audience: Ages 7-10 | Audience: Grades 2-3 | Summary: "Full-color photographs and a running timeline illustrate important developments in the history of telephones. Topics feature the telegraph, Alexander Graham Bell, mobile and cellular phones, satellites, text messaging, and video calling. A table of contents, glossary, further resources, and an index are included"-- Provided by publisher.
Identifiers: LCCN 2019036333 (print) | LCCN 2019036334 (ebook) | ISBN 9781681519623 (library binding) | ISBN 9781681526096 (paperback) | ISBN 9781645490470 (pdf)
Subjects: LCSH: Telephone--History--Juvenile literature.
Classification: LCC TK6165 .S65 2022 (print) | LCC TK6165 (ebook) | DDC 621.38609--dc23
LC record available at https://lccn.loc.gov/2019036333
LC ebook record available at https://lccn.loc.gov/2019036334

Editor: Gillia Olson
Designer: Aubrey Harper
Photo Researcher: Bridget Prehn

Photo Credits: Alamy/Alex Segre 23, North Wind Picture Archives 4, PA Images 28–29; AP/Associated Press 15, Eric Rissberg 18–19; Getty/David Paul Morris 24–25; iStock/doug4537 9; iStock/KOHb circuit texture (throughout), mikimad 27 (phones), PhotoTalk 27 (woman), SDI Productions 27 (couple), umesh chandra cover (back), Library of Congress/Gilbert H. Grosvenor Collection of Photographs of the Alexander Graham Bell Family 6–7, Van Epps, John 8; Shutterstock/alexandarilich 10–11, gfdunt 14, Lawrence Roberg 16–17, marianstock cover (front), Monkey Business Images 20; University of Wisconsin-Milwaukee Libraries/James Blair Murdoch Photographs 12–13

TABLE OF CONTENTS

America on the Wire 5

Reaching Out 9

Connecting Farther and Faster 13

Going Mobile 22

Glossary 30

Read More 31

Websites 31

Index 32

Samuel Morse invented the telegraph in 1844.

Samuel Morse invents the telegraph.

1844

. . L O A D I N G . . . L O A D I N G .

America on the Wire

In 1825, Samuel Morse got a letter. His wife was very sick. The letter got to him too late. By the time he got home, she had died. Morse looked for a way to send messages faster. In 1844, he invented the **telegraph**. It sent **electricity** along a wire. He made a code of long and short beeps. The beeps stood for letters. Morse code messages took just minutes.

OADING . . . LOADING . . . LOADING . . .

It took another 30 years to send voices over wires. Alexander Graham Bell, Elisha Gray, and others worked on it. Bell put in a **patent** for his telephone design on February 14, 1876. He beat Gray by a few hours. A few weeks later, on March 10, Bell made the first phone call. It was to his assistant. He said, "Mr. Watson, come here. I want to see you."

By 1892, Bell was making the first phone call between New York and Chicago.

Samuel Morse invents the telegraph.

1844

MARCH 10, 1876

Bell makes the first phone call.

LOADING . . . LOADING . . .

OADING . . . LOADING . . . LOADING . . .

Telephone operators work at a switchboard in Hamburg, New York, about 1910.

Samuel Morse invents the telegraph.

The first telephone exchange opens.

1844

MARCH 10, 1876

1878

G . . . LOADING .

Bell makes the first phone call.

Reaching Out

Telephones needed wires to connect them. In 1878, the first telephone **exchange** was built. Phone wires went from houses to the exchange. A person, the **operator**, connected each call by hand.

Almon Strowger thought machines should connect calls, not operators. He invented an automatic telephone exchange in 1889. Then he made a rotary phone to dial phone numbers. People spun a dial with numbers to make a call.

LOADING... LOADING... LOADING...

Early phones had separate mouth and earpieces. The candlestick phone had a mouthpiece on a small stand. The earpiece was connected by a wire. In 1904, the Bell Company made its first French Phone. It had the ear and mouthpiece on one handset.

The French Phone had the earpiece and mouthpiece on one handset.

- 1844: Samuel Morse invents the telegraph.
- MARCH 10, 1876: Bell makes the first phone call.
- 1878: The first telephone exchange opens.
- 1904: Bell Company introduces the first French Phone.

OADING . . . LOADING . . . LOADING . . .

A man uses a car phone in 1956.

1844	Samuel Morse invents the telegraph.
MARCH 10, 1876	Bell makes the first phone call.
1878	The first telephone exchange opens.
1904	Bell Company introduces the first French Phone.
1946	The first mobile call is made using a car phone.

Connecting Farther and Faster

The first car phone call was made in 1946. These early **mobile** phones used **radio** signals. The phones were big and costly. They had to be carried in cars or trucks. People often had to wait to make calls. They needed an open radio channel.

OADING . . . LOADING . . . LOADING . . .

The 1950s and '60s brought calling across the world. In 1956, a big wire cable was laid along the bottom of the Atlantic Ocean. It ran from Canada to Scotland. People could call across the sea. In 1962, NASA launched Telstar I. It was the first communications **satellite**. Telstar could bounce phone signals around the world. Today at least 800 similar satellites orbit Earth.

Telstar 1

- **1844** Samuel Morse invents the telegraph.
- **MARCH 10, 1876** Bell makes the first phone call.
- **1878** The first telephone exchange opens.
- **1904** Bell Company introduces the first French Phone.
- **1946** The first mobile call is made using a car phone.
- **1956** The first working transatlantic cable is finished.

The first cable to cross the ocean comes ashore in Canada in 1956.

OADING . . . LOADING . . . LOADING . . .

More than 70 years after the rotary dial was invented, phones got buttons. In 1963, phone company AT&T introduced touch-tone phones. The caller pressed buttons to enter a phone number. The phone sent tones along the wire to make connections. By the 1990s, touch-tone phones were more common than rotary ones.

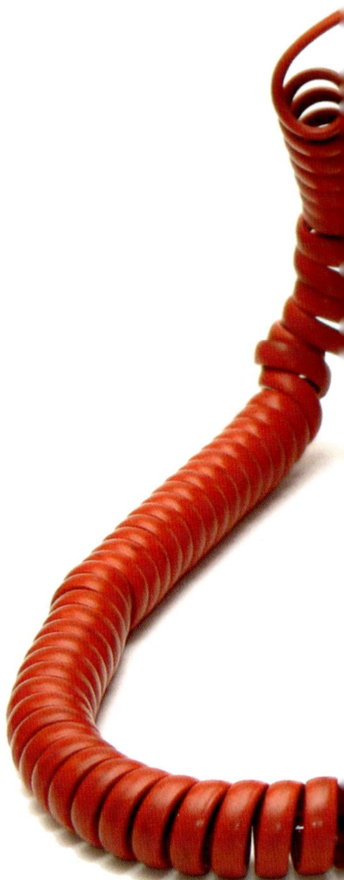

The touch-tone phone also had buttons for "star" and "pound" used for special calling features.

- 1844: Samuel Morse invents the telegraph.
- March 10, 1876: Bell makes the first phone call.
- 1878: The first telephone exchange opens.
- 1904: Bell Company introduces the first French Phone.
- 1946: The first mobile call is made using a car phone.
- 1956: The first working transatlantic cable is finished.

LOADING . . . LOADING . . .

1963

AT&T introduces the first touch-tone phone.

On April 3, 1973, Martin Cooper made the first **cellular** phone call. It was on a Motorola Dynatac. Ten years later, this phone would become the first cell phone sold in stores. Cell towers were built all over. They sent call signals back and forth. Each tower covered an area called a cell. All these towers together became the cellular network.

Martin Cooper, shown in 2003, poses with the Motorola Dynatac he used in 1973.

- 1844 — Samuel Morse invents the telegraph.
- MARCH 10, 1876 — Bell makes the first phone call.
- 1878 — The first telephone exchange opens.
- 1904 — Bell Company introduces the first French Phone.
- 1946 — The first mobile call is made using a car phone.
- 1956 — The first working transatlantic cable is finished.

Motorola designer Martin Cooper makes the first cell phone call.

1963 1973 ING . . . LOADING . . .

AT&T introduces the first touch-tone phone.

Cordless phones let you move around your house while taking a call.

- **1844** — Samuel Morse invents the telegraph.
- **MARCH 10, 1876** — Bell makes the first phone call.
- **1878** — The first telephone exchange opens.
- **1904** — Bell Company introduces the first French Phone.
- **1946** — The first mobile call is made using a car phone.
- **1956** — The first working transatlantic cable is finished.

Before cell phones took over, people got cordless phones at home. The cordless phone was patented in 1977. Cordless phones have a base and a handset. The base is plugged into the phone line. Calls go over radio waves between the base and handset. People were no longer tied to their phone by a cord. As they improved, callers could go farther and farther from the base.

1963 — AT&T introduces the first touch-tone phone.

1973 — Motorola designer Martin Cooper makes the first cell phone call.

1977 — The modern cordless phone is patented.

. . . LOADING . . .

Going Mobile

More and more people got cell phones. But what about times you couldn't answer the phone? Or you just needed a quick answer? Enter text messaging. On December 3, 1992, Neil Papworth sent the first text message. It said "Merry Christmas." In 1993, Nokia sold the first cell phone with full text messaging.

Nokia phones with text messaging were big sellers in the 1990s.

1844: Samuel Morse invents the telegraph.

MARCH 10, 1876: Bell makes the first phone call.

1878: The first telephone exchange opens.

1904: Bell Company introduces the first French Phone.

1946: The first mobile call is made using a car phone.

1956: The first working transatlantic cable is finished.

1963 — AT&T introduces the first touch-tone phone.

1973 — Motorola designer Martin Cooper makes the first cell phone call.

1977 — The modern cordless phone is patented.

DECEMBER 3, 1992 — The first text message is sent.

ADING . . .

- **1844** — Samuel Morse invents the telegraph.
- **MARCH 10, 1876** — Bell makes the first phone call.
- **1878** — The first telephone exchange opens.
- **1904** — Bell Company introduces the first French Phone.
- **1946** — The first mobile call is made using a car phone.
- **1956** — The first working transatlantic cable is finished.

In the 2000s, cell phones turned into smart phones. Sharp made the first cell phone that could send pictures in 2000. Other phones began to have email and internet access. In 2007, Apple introduced the iPhone. It was a small touch-screen computer that made calls. It had a camera, media player, and ran **apps**. The iPhone put a computer in people's pockets.

Apple CEO Steve Jobs debuts the iPhone in 2007.

- **1963** AT&T introduces the first touch-tone phone.
- **1973** Motorola designer Martin Cooper makes the first cell phone call.
- **1977** The modern cordless phone is patented.
- **DECEMBER 3, 1992** The first text message is sent.
- **JANUARY 2007** Steve Jobs introduces the iPhone. It goes on sale a few months later.

For years, people had wanted to see the person they called. The Picturephone came out in 1964. But it didn't catch on. In the 2000s, internet service got faster. People video chatted through their computers. In 2010, Apple put Facetime video calling on the iPhone. By 2018, WhatsApp users made 55 million video calls a day.

Video calling became popular in the 2010s.

1844	Samuel Morse invents the telegraph.
MARCH 10, 1876	Bell makes the first phone call.
1878	The first telephone exchange opens.
1904	Bell Company introduces the first French Phone.
1946	The first mobile call is made using a car phone.
1956	The first working transatlantic cable is finished.

1963	AT&T introduces the first touch-tone phone.
1973	Motorola designer Martin Cooper makes the first cell phone call.
1977	The modern cordless phone is patented.
DECEMBER 3, 1992	The first text message is sent.
JANUARY 2007	Steve Jobs introduces the iPhone. It goes on sale a few months later.
2010	Facetime is available for iPhone 4.

1844	March 10, 1876	1878	1904	1946	1956
Samuel Morse invents the telegraph.	Bell makes the first phone call.	The first telephone exchange opens.	Bell Company introduces the first French Phone.	The first mobile call is made using a car phone.	The first working transatlantic cable is finished.

In the future, we might be able to make holographic video calls. A **hologram** is a 3-D picture shown like a movie. Right now, we need special equipment to make them. Soon, our phones may be able to do it. Calling people will almost bring them into the room with you!

UK soccer star Steph Houghton calls a young fan during the first 5G holographic phone call.

- 1963: AT&T introduces the first touch-tone phone.
- 1973: Motorola designer Martin Cooper makes the first cell phone call.
- 1977: The modern cordless phone is patented.
- DECEMBER 3, 1992: The first text message is sent.
- JANUARY 2007: Steve Jobs introduces the iPhone. It goes on sale a few months later.
- 2010: Facetime is available for iPhone 4.

Glossary

app A program used on a computer; "app" stands for application.

cellular Relating to a system that uses radio waves instead of wires to send telephone signals.

electricity A form of energy that is carried through wires.

exchange A central place where phone calls are connected.

hologram A three-dimensional (3-D) picture made by a complex pattern of light.

mobile Able to move from one place to another.

operator A person who connects phone calls.

patent An official document giving an inventor the only right to make, use, or sell his or her invention for a period of time.

radio The use of invisible waves to send sound that has been turned into electrical energy.

satellite A machine sent into space that circles the earth, moon, or other object.

telegraph A way of sending messages as electrical signals over wires.

Read More

Coates, Eileen S. *Alexander Graham Bell's Telephone.* STEM Milestones: Historic Inventions and Discoveries. New York: PowerKids Press, 2019.

Colby, Jennifer. *Telephone to Smartphones.* Then to Now Tech. Ann Arbor, Mich.: Cherry Lake Publishing, 2019.

Fraser, Mary Ann. *Alexander Graham Bell Answers the Call.* Watertown, Mass.: Charlesbridge, 2017.

Websites

AT&T History: See How People Learned to Use Dial Telephones

https://thekidshouldseethis.com/post/introducing-the-dial-telephone-films-from-1936-1954

DK Find Out: Alexander Graham Bell

www.dkfindout.com/us/science/famous-scientists/alexander-graham-bell

Library of Congress: Who Invented the Telephone?

www.loc.gov/rr/scitech/mysteries/telephone.html

Index

Bell, Alexander Graham 6
candlestick phone 10
car phone 13
cellular network 18
cellular phones 18, 21, 22, 25
Cooper, Martin 18
cordless phones 21
exchanges 9
Facetime 26
French phone 10
Gray, Elisha 6
holograms 29
iPhone 25, 26
mobile phones.
See cellular phones
Morse, Samuel 5
operators 9
patents 6
radio signals 13
rotary phones 9, 16
satellites 14
smart phones 25
Strowger, Almon 9
telegraph 5
Telstar I 14
text messaging 22
touch-tone phones 16
transatlantic cable 14
video calls 26, 29
WhatsApp 26
wires, telephone 9, 14

About the Author

Angie Smibert has written dozens of educational titles just like this one. She was a science writer and online training developer at NASA's Kennedy Space Center for many, many years. She received NASA's prestigious Silver Snoopy as well as several other awards for her work.